TECHNICAL
REPORT

United States Air Force Aircraft Fleet Retention Trends

A Historical Analysis

Timothy L. Ramey, Edward G. Keating

Prepared for the United States Air Force

The research described in this report was sponsored by the United States Air Force under Contract FA7014-06-C-0001. Further information may be obtained from the Strategic Planning Division, Directorate of Plans, Hq USAF.

Library of Congress Cataloging-in-Publication Data

Ramey, Timothy L., 1951–
United States Air Force aircraft fleet retention trends : a historical analysis / Timothy L. Ramey, Edward G. Keating.
p. cm.
Includes bibliographical references.
ISBN 978-0-8330-4794-6 (pbk. : alk. paper)
1. United States. Air Force—Equipment—History. 2. Airplanes, Military—United States—Design and construction—History. I. Keating, Edward G. (Edward Geoffrey), 1965– II. Title.

UG1243.R363 2009
358.4'183—dc22

2009044164

Published 2009 by the RAND Corporation
1776 Main Street, P.O. Box 2138, Santa Monica, CA 90407-2138
1200 South Hayes Street, Arlington, VA 22202-5050
4570 Fifth Avenue, Suite 600, Pittsburgh, PA 15213-2665
RAND URL: http://www.rand.org/
To order RAND documents or to obtain additional information, contact
Distribution Services: Telephone: (310) 451-7002;
Fax: (310) 451-6915; Email: order@rand.org

Preface

An extensive body of literature highlights the challenges of aging aircraft. The basic concern is that aircraft become more expensive to maintain and less available for operations as they age. Dixon (2006) presents a review of this literature.

In this technical report, we do not directly examine the age of the United States Air Force's (USAF's) aircraft. Instead, we provide historical, contextual information on the ages of aircraft designs operated by the USAF. Using two 1998 Air Force Historical Agency reports, the report traces the use of different aircraft designs in the Air Force and its predecessor organizations, dating back to the earliest days of military aviation through 1995.

Since the end of World War II and the formation of the Air Force as an independent military service in 1947, there has been a secular trend for the Air Force to keep aircraft designs in operation for ever-longer periods. So, while the mean age of aircraft designs currently in operation is at an all-time high, the same statement could have been made at most times throughout the history of the Air Force. The Air Force has had, by and large, an ever-aging portfolio of designs.

In theory, the Air Force could have a new aircraft that was manufactured using an old design. With the exception of a handful of designs such as the C-130, however, Air Force aircraft have typically been only a few years younger than the design from which they were manufactured.

This report is not designed to have direct policy implications. Instead, it is a contribution to a body of factual knowledge about aging aircraft and the challenges that the Air Force and the Department of Defense face. Related RAND Corporation documents include the following:

- *Aging Aircraft: USAF Workload and Material Consumption Life Cycle Patterns*, by Raymond A. Pyles (MR-1641-AF), 2003.
- *The Maintenance Costs of Aging Aircraft: Insights from Commercial Aviation*, by Matthew Dixon (MG-486-AF), 2006.

This research is intended to be of interest to Air Force and other Department of Defense personnel involved with aircraft fleet management. The research was conducted within the Resource Management Program of RAND Project AIR FORCE in the context of aging aircraft research sponsored by the Director of Strategic Planning, Deputy Chief of Staff for Strategic Plans and Programs (AF/A8X).

RAND Project AIR FORCE

RAND Project AIR FORCE (PAF), a division of the RAND Corporation, is the U.S. Air Force's federally funded research and development center for studies and analyses. PAF provides the Air Force with independent analyses of policy alternatives affecting the development, employment, combat readiness, and support of current and future aerospace forces. Research is conducted in four programs: Force Modernization and Employment; Manpower, Personnel, and Training; Resource Management; and Strategy and Doctrine.

Additional information about PAF is available on our Web site:
http://www.rand.org/paf/

Contents

Figures

Tables

Summary

In this report, we review the history of the Air Force and its predecessor organizations and find that, since the founding of the Air Force in 1947, the service has routinely had fleets of aircraft with older designs than it had ever operated previously.

In the report, we do not directly examine the age of Air Force aircraft. Instead, we analyze the age of aircraft designs in use by the Air Force. In general, Air Force aircraft have been only a few years younger than the designs from which they were manufactured.

Historical Sources

The primary sources for this historical analysis are two reports published in 1998 by the Air Force Historical Agency: *USAF Active Flying, Space, and Missile Squadrons as of 1 October 1995* and *Active Air Force Wings as of 1 October 1995*.

Using these reports, we look across active wings and squadrons to identify the first year in which any active wing or squadron reported operating a specific aircraft design. In a similar manner, we identify the last year in which a given aircraft design was reported as being operated by any active wing or squadron. The data do not cover Air Force Reserve or Air National Guard units. (See pp. 3–4.)

The historical reports show a ramp-up in the number of aircraft designs in operation in the late 1910s into the 1920s, as military aviation took hold. Since roughly 1930, there have generally been 30–40 different aircraft designs in operation at any one time. (See pp. 5–6.)

In the early years of military aviation and during World War II, a large number of aircraft designs were introduced and many designs were retired. The last large-scale introduction of new designs was during the Vietnam War. Since then, the Air Force's fleet of aircraft has been generally stable, with a few design introductions and retirements in a typical year. (See pp 5–6.)

Patterns in Aircraft Design Age

There have been nearly consistent increases in maximum, median, and mean operated aircraft design age. There has also been a long-term increase in the dispersion of ages of aircraft designs in operation. (See pp. 7–8.)

This analysis finds that the age of aircraft designs in their last year of operation has trended up. Designs survived roughly five years in the earliest years of the Air Force. Designs retired in the 1990s, by contrast, had been in operation an average of roughly 20 years. The growth in retirement ages is relatively consistent and constant. (See pp. 10–12.)

The periods with the most introductions of new designs (1916–1920 and 1941–1945) also had the highest percentage of introductions that were operated for five or fewer years. During periods with many introductions, substantial experimentation appears to have been occurring. By contrast, more-recent periods have seen longer-lasting designs and relatively fewer short-lived designs. (See pp. 11–12.)

This report shows that, since the end of World War II and the formation of the Air Force as an independent military service, there has been a pronounced trend for the Air Force to keep aircraft designs in operation for ever-longer periods. Our results do not speak to why or how this pattern of aging has occurred. They do suggest, however, that the pattern is both consistent and persistent. Therefore, while the mean age of aircraft designs currently in operation is at an all-time high, the same statement could have been made at most times throughout the history of the Air Force. By and large, the Air Force has operated an ever-aging portfolio of aircraft designs. (See pp. 15–16.)

Acknowledgments

Richard Moore and Kelvin Utendorf of the Air Force Materiel Command provided helpful comments on an earlier draft of this document, as did RAND colleagues Laura Baldwin, Jean Gebman, and the late Hyman L. Shulman. Daniel Norton suggested relevant literature. During the initial research for this survey, the support of USAF Lt Col John Orsato was most helpful. Suggestions, criticism, and encouragement from Ray Pyles are gratefully acknowledged. Jane Siegel helped prepare, and Miriam Polon edited, this report.

We received reviews of an earlier version of this paper from Mark Lorell of RAND and RAND adjunct researcher Michael W. Wynne, the 21st Secretary of the Air Force. Cynthia Cook and Carl Rhodes coordinated the PAF quality assurance process. We also thank Philip Antón, Lt Brandon Dues, and Deborah Peetz for their comments at a RAND payday logistics seminar held on June 25, 2009.

Remaining errors are, of course, the authors' responsibility.

Abbreviations

AAS	*USAF Active Flying, Space, and Missile Squadrons as of 1 October 1995*
AAW	*Active Air Force Wings as of 1 October 1995*
CSAF	Chief of Staff of the United States Air Force
PAF	RAND Project AIR FORCE
UAV	unmanned aerial vehicle
USAF	United States Air Force

CHAPTER ONE

Introduction

> The B-52 is going to fall apart on us before we can get a replacement for it. There is serious danger that this may happen to us.

This statement, made by a Chief of Staff of the United States Air Force (CSAF), illustrates the challenge posed by aging aircraft. The Air Force is operating aircraft that are increasingly aged. Some observers fear that older aircraft will be more expensive to maintain because of such challenges as diminishing manufacturing sources and the longer times required for repairs. Because of long lags in the aircraft procurement process, fleet replacement can take many years. Thus, legacy aircraft will continue to be operated long after a replacement program commences.

Of special interest to this report is the provenance of the opening quote presented above. Made in 1964, the statement is from then-CSAF Gen Curtis E. LeMay.[1] What we now term the "aging aircraft issue" has been challenging the Air Force for decades. General LeMay was making the argument to replace the B-52 based on concerns about aging aircraft in 1964. Yet B-52s remain in the Air Force's inventory today.

This report reviews the history of the Air Force and its predecessor organizations and finds that, since the founding of the Air Force in 1947, the service has routinely had aircraft designs in operation that were older than the service had ever observed to that date.[2] Aircraft designs in operation in the 1960s were generally older than those of the 1950s; aircraft designs in operation in the 1970s were generally older than those of the 1960s; and so forth. While our historical analysis extends only through the mid-1990s, a similar pattern has continued through the present. By and large, the Air Force has had ever-aging aircraft designs in operation since its inception as an independent military service.

This report reviews the history of fixed-wing aircraft designs operated by the Air Force. This report does not speak to why or how designs were kept in operation for the durations observed. Many factors—economic, technological, and political—may have been involved. We know, for example, that the focus of the aircraft design process has changed over the generations, from aerodynamics to aeronautics to avionics to stealth. In parallel, the Air Force has

---

[1] This quote is reported in Worden, 1998, who cites Futrell, 1989.

[2] The U.S. Air Force was founded September 18, 1947. It was preceded by the Aeronautical Division, U.S. Signal Corps (August 1, 1907–July 18, 1914); the Aviation Section, U.S. Signal Corps (July 18, 1914–May 20, 1918); the Division of Military Aeronautics (May 20, 1918–May 24, 1918); the U.S. Army Air Service (May 24, 1918–July 2, 1926), the U.S. Army Air Corps (July 2, 1926–June 20, 1941); and the U.S. Army Air Forces (June 20, 1941–September 18, 1947). See Air Force Historical Studies Office, 2009.

made investments in durability technologies to enable longer service lives. However, this evolution in design philosophy and its possible effects on design longevity are beyond the scope of this report.

In this report, we do not directly examine the age of Air Force aircraft. Instead, we analyze the age of aircraft designs in use by the Air Force. Although there are a handful of cases of the Air Force buying new aircraft derived from old designs (the venerable C-130 is the best such example), Air Force aircraft have generally been only a few years younger than the designs from which they were manufactured.

This report is not designed to have direct policy implications. Instead, it is a contribution to a body of factual knowledge about aging aircraft and the challenges that the Air Force and Department of Defense face.

The remainder of this report is structured as follows: Chapter Two discusses the two Air Force Historical Research Agency reports that undergird our analysis. Chapter Three presents patterns in aircraft design age found in the data. Chapter Four provides a concluding discussion. Appendix A presents tables identifying the aircraft designs utilized in this analysis, and Appendix B provides evidence that our findings are not meaningfully different, whether or not one includes "one-report" designs in the analysis.

CHAPTER TWO

Historical Sources

The primary sources for this historical analysis are two reports published in 1998 by the Air Force Historical Research Agency: *USAF Active Flying, Space, and Missile Squadrons as of 1 October 1995* and *Active Air Force Wings as of 1 October 1995* (hereafter abbreviated AAS and AAW, respectively).

The compiled histories recount the lineage, assignments, component organizations, stations, commanders, aircraft operated, operations participated in, honors, decorations, and emblems of each organization. Included in the histories are statements that identify the aircraft that were operated by those wings and squadrons and the periods of time over which those aircraft were operated.

Table 2.1 provides an illustrative example of aircraft operation as reported in AAW's entry for the 15th Air Base Wing. (There is nothing distinctive about the 15th Air Base Wing. We are using it only as an example to illustrate what AAW data look like.) The structure of the information provided by AAS is similar, but it covers squadrons rather than wings.

Although the information shown in Table 2.1 for the 15th Air Base Wing goes back only to 1940, AAW includes reports of aircraft that were operated as early as 1932. AAW covers

Table 2.1
Aircraft Operated by the 15th Air Base Wing

Aircraft	Years	Aircraft	Years
A-12	1940–1942	P-61	1944
OA-9	1940–1942	A-26	1946
P-26	1940–1942	F-86	1955–1958
P-36	1940–1942	F-102	1958–1960
B-12	1941–1942	F-84	1962–1964
P-39	1941–1944	T-33	1962–1970 1972–1987
P-40	1941–1944	F-4	1964–1970
P-47	1943–1946	B-57	1968–1970
P-70	1943–1944	EC-135	1971–1992
A-24	1944	O-2	1972–1980
P-51	1944–1946	C-135	1992–1995

SOURCE: AAW.

87 different wings over 64 years, with the maximum number of wings in existence at a time being 70 during the Vietnam War era. More extensive in its coverage, AAS includes reports of aircraft that were in operation as early as 1913. Records prior to 1947 cover the Air Force's various predecessor organizations. AAS covers 257 squadrons over 83 years. The peak number of squadrons extant in a single year was just over 200 in 1943.

For each aircraft design we identified in AAS and AAW (e.g., the F-86), we looked across squadrons (in AAS) and wings (in AAW) to identify the first year in which any active duty wing or squadron reported operating that aircraft design. We then located the last year in which that design was reported as being operated by any active duty wing or squadron. In the analysis reported here, we interpret the difference between the earliest operation reported anywhere in AAS or AAW and the latest reported operation to be the effective service life of that aircraft design. Any design not in operation in 1995, according to AAS or AAW, is assumed to have been retired before that year.

The aircraft designs used in our analysis are tabulated in Appendix A. Some reported designations have been consolidated under a common design "parent." For example, the DB-7, F-3, and P-70 designations are all variants of the A-20 design. On other occasions, the same designation was used for multiple designs. For example, a design variously called the A(B)-26, A/RA-26, A-26, FA(RB)-26, or FA-26 is a design distinct from the one variously called the B/RB-26, B-26, RB-26, or WB-26. We observed 4,026 reports of designs operated under 580 distinct designations. These represent 325 distinct air vehicle designs, of which 282 were for air-breathing aircraft—those examined in this report. We do not examine balloons, blimps, helicopters, missiles, or unmanned aerial vehicles (UAVs).

There is an element of judgment associated with tabulating unique aircraft designs. To some extent, we were constrained by AAS/AAW decisions (e.g., they do not differentiate F-16 variants). We made an effort to identify "designs" at a consistent level of detail across the database, being sensitive to the amount of detail we could find in the sources available to us. We do not believe any of the broad findings in this report are sensitive to our design categorization decisions. In Appendix B, for instance, we reprise some of our figures, removing aircraft designs reported only once in AAS and AAW. The major results are essentially unchanged.

For purposes of this analysis, AAS and AAW have some shortcomings in common. Any aircraft design that was never operated in one of the active wings or squadrons covered by AAS and AAW would not appear in our analysis. Because AAS and AAW do not cover Air Force Reserve or Air National Guard units, any fleet that was moved solely into the Guard or Reserve would be seen as retired in the year it was transferred. Of course, because the AAS and AAW data go only through 1995, introductions or retirements subsequent to that year are not considered in this analysis.

Aircraft design age is not the same as aircraft age. Of our 282 air vehicle designs, we were able to find 163 of them in Baugher's (2008) database. Baugher's data show the oldest (first produced) and newest (last produced) aircraft of a given design. The Baugher data we analyzed ran through 1999, so that was the latest year in which we would see an aircraft produced. We define a design's production duration to be 1 plus the last production year minus the first production year. Table 2.2 shows the distribution of production run durations according to these data.

Our production run duration estimation methodology applied to the Baugher data suggests that the plurality of aircraft designs were produced in a single year. A handful of designs, shown at the bottom of Table 2.2, were so successful that they were produced over longer

Table 2.2
Estimated Production Run Durations

Duration (Years)	Aircraft Designs	Design	First Year–Last Year
1	44		
2	21		
3	15		
4	30		
5	18		
6	10		
7	5		
8	1		
9	3		
10	6		
11	2		
12	2		
15	1	A-7	1967–1981
17	1	F-4	1962–1978
22	1	F-16	1978–1999
24	1	F-5	1963–1986
28	1	F-15[a]	1971–1998
47	1	C-130	1953–1999

SOURCE: Baugher, 2008.

[a] The database does not differentiate F-15Es from earlier F-15 variants.

periods. For those designs, in particular, our analysis is too pessimistic, i.e., many aircraft are considerably younger than their designs.

Utilizing the AAS and AAW data, Figure 2.1 illustrates the number of aircraft designs in operation in each year. Only fixed-wing, air-breathing Air Force aircraft (e.g., bombers, cargo aircraft, fighters) are considered in this report. Data on helicopters, lighter-than-air vehicles, and missiles are also available in AAS and AAW but are not reflected in this analysis.

As shown in Figure 2.1, there was a ramp-up in the number of aircraft designs in operation in the late 1910s into the 1920s as military aviation took hold. Since about 1930, there have generally been 30–40 different aircraft designs in operation at any point in time. The exceptions have been surges in the number of aircraft designs in operation in World War II and during the Vietnam War. Note that Figure 2.1 portrays the number of different aircraft *designs* in operation, not the number of aircraft. The number of new aircraft designs entering operation and the number of existing aircraft designs retired each year is shown in Figure 2.2.

In the early years of military aviation and again during World War II, many new aircraft designs were introduced and many were also retired. The last large-scale introduction of designs was during the Vietnam War. Since then, the Air Force's fleet of aircraft has been generally stable, with few or no design introductions and retirements each year.

Figure 2.1
Aircraft Designs in Operation, by Year

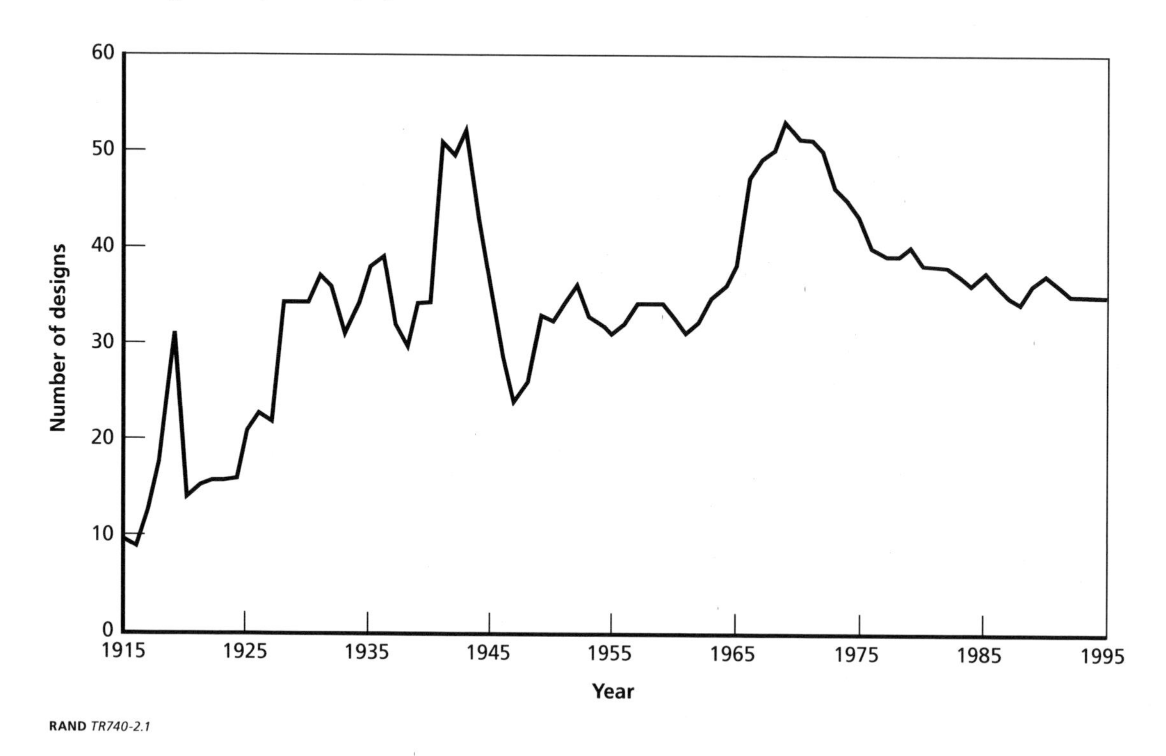

RAND *TR740-2.1*

Figure 2.2
Introductions and Retirements of Aircraft Designs, by Year

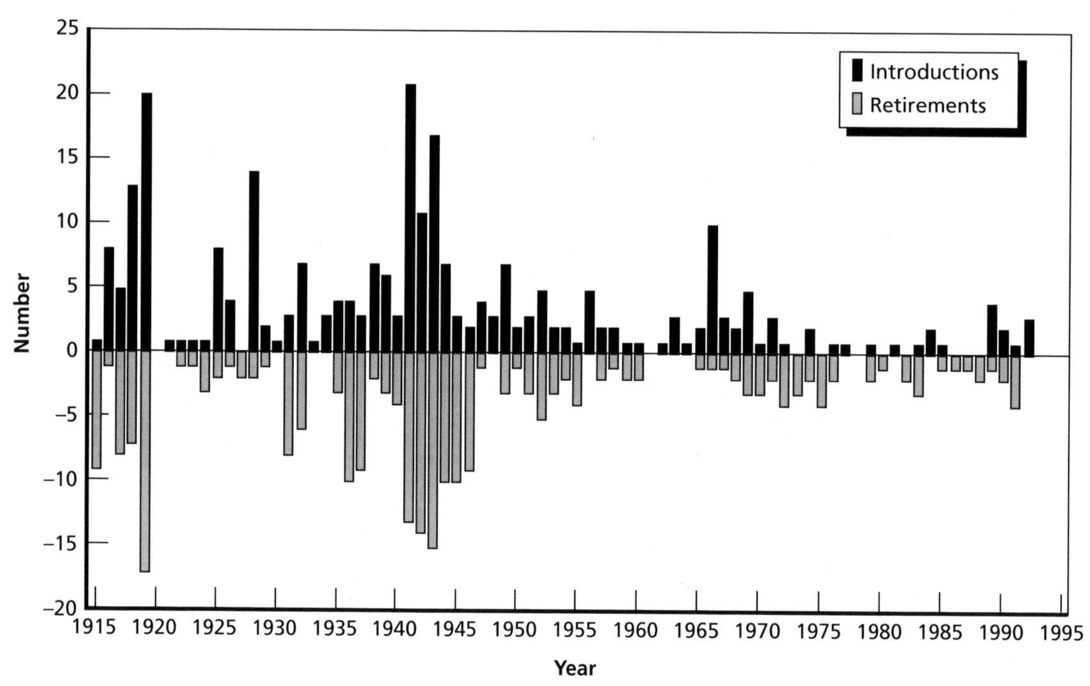

RAND *TR740-2.2*

CHAPTER THREE

Patterns in Aircraft Design Age

Figure 3.1 shows the maximum age of designs in operation each year from 1915 to 1995, the median design age in each year, and the mean age of the designs in operation each year.[1] Between 1950 and 1995, there were only seven years in which the mean age of designs in operation did not increase. Two of those seven years were during the Vietnam War (1966 and 1969). The others were 1956 (the B-17 was retired and the C-130 introduced), 1976 (the C-47 was retired and the C-12 introduced), 1984 (the F-100 was retired and the C-21 introduced), 1989 (the B-2 and F-22 were introduced), and 1990 (the F-5 was retired and the C-25 introduced). The mean age of a design in operation increased from 5.7 years in 1950 to 21.9 years

Figure 3.1
Maximum, Median, and Mean Age of Designs in Operation, by Year

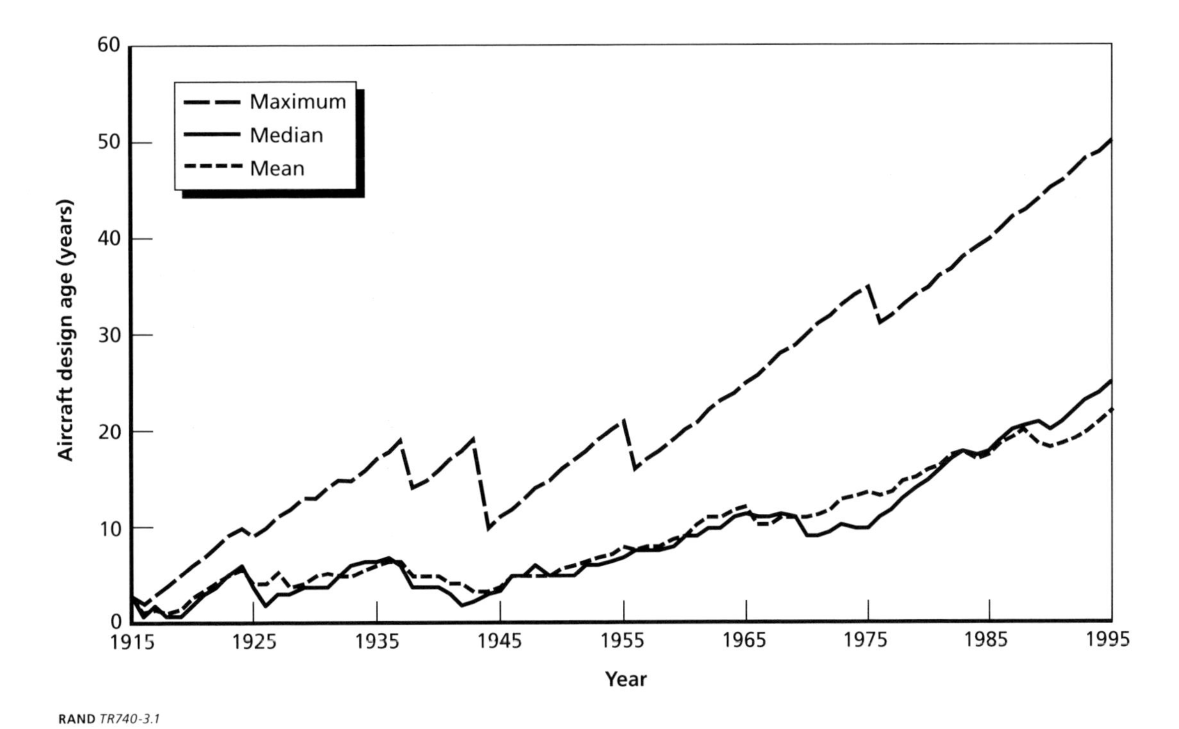

RAND *TR740-3.1*

[1] We define an aircraft in its first year of operation to be one year old, in its second year to be two years old, etc. This tally technique is contrary to the "birthday" technique whereby, for instance, someone's 21st birthday represents the entrance into his or her 22nd year of life.

in 1995.[2] Since 1944, the longest-lasting designs have been the B-17 (introduced in 1935 and operated until 1955), the C-47 (introduced in 1941 and operated until 1975), and the B-52 (introduced in 1946 and still in operation today).

As the maximum, median, and mean ages of aircraft designs in operation have increased, so too has the standard deviation of the ages of designs in operation. Figure 3.2 plots the standard deviation in the ages of operated designs by year. Not only have Air Force maintainers faced older designs, they have also confronted growing dispersion in the age of designs they maintain.

The Baugher data suggest that production runs have increased in length over time—from a couple of years in the 1930s to nearly 10 years in the 1990s. Consequently, there has been a tendency for an increasing number of aircraft to be younger than their designs. To the extent that this is so, Figure 3.1's pattern is overly pessimistic because more recent aircraft are a number of years newer than their designs.

In terms of Figure 3.1's oldest post–World War II aircraft, the Baugher data indicate that the B-17 had a seven-year production run (1938–1944), the C-47 had a five-year production

Figure 3.2
Standard Deviation of Age of Designs in Operation, by Year

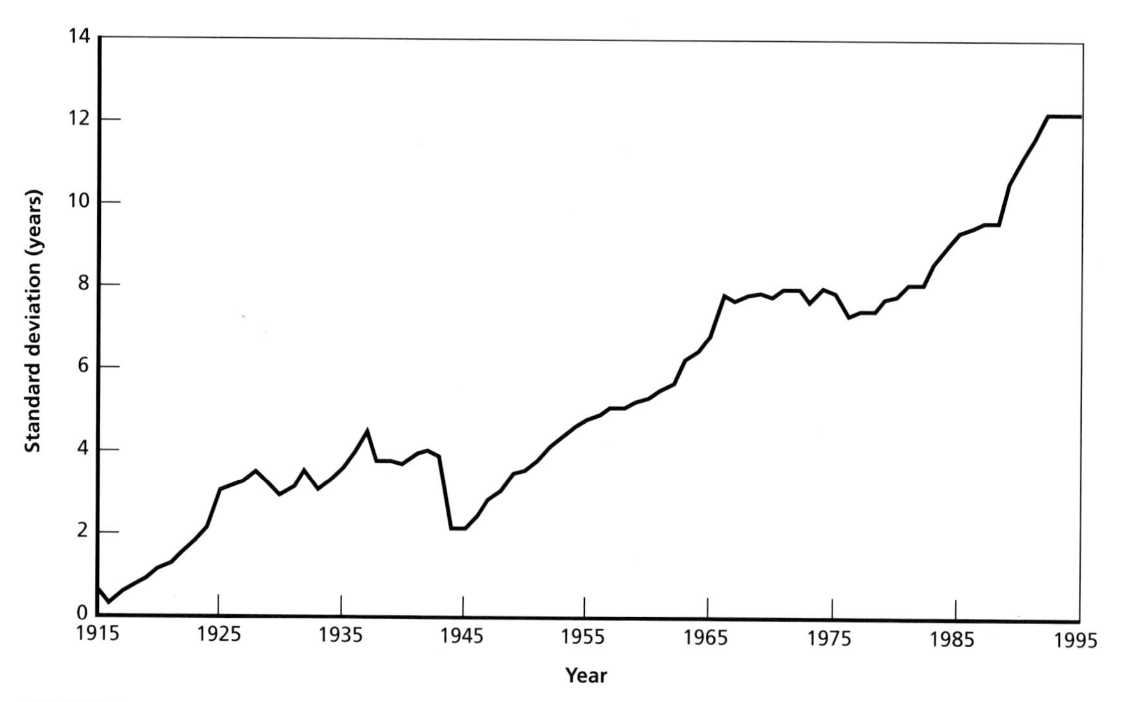

RAND *TR740-3.2*

2 This mean age is not weighted by the number of aircraft in each fleet. In 1995, for example, there were 35 fleets in operation. They ranged in age from 4 (the F-15E, the T-1, and the T-43) to 50 (the B-52). The arithmetic average of these 35 fleets' design ages was 21.9 years. The average fleet age would, however, be different because each fleet has a different number of aircraft in it.

run (1941–1945), and the B-52 had a 10-year production run (1952–1961).[3] Longer production runs do not, however, appear to explain the general growth in design longevity.

Fleet size also affects the mean age of *aircraft* in operation, Air Force–wide. For example, if the largest fleets happened to be of newer designs, the mean age of aircraft would be less than our design-level statistics indicate. (We do not address trends in the average age of aircraft in operation.)

Table 3.1 presents the same data as shown in Figure 3.1, but aggregated into five-year blocks. Since the end of World War II, there has been a nearly continuous increase in the maximum, median, and mean design age of aircraft in operation. The mean design age dropped slightly during the Vietnam War, which was the occasion of the last large-scale introduction of new aircraft designs into the Air Force. The five-year block pattern for all three statistics has been increasing monotonically since 1976.

Table 3.1
Maximum, Median, and Mean Age of Designs in Operation, in Five-Year Blocks

Time Period	Maximum	Median	Mean
1916–1920	6	1	1.7
1921–1925	10	4.5	4.5
1926–1930	13	3	4.5
1931–1935	17	5	5.6
1936–1940	19	5	5.7
1941–1945	19	3	3.9
1946–1950	16	5	5.2
1951–1955	21	6	7.0
1956–1960	20	8	8.4
1961–1965	25	10	11.2
1966–1970	30	11	10.9
1971–1975	35	10	12.6
1976–1980	35	13	14.7
1981–1985	40	17	17.3
1986–1990	45	20	19.0
1991–1995	50	22	20.0

[3] As shown in Table A.2, however, the AAS/AAW tabulation first observes the B-52 in 1946. Boeing agrees with Baugher, indicating the B-52 first flew in 1952 (Boeing, 2002). However, contrary to Baugher, Boeing suggests the B-52 had an 11-year production run from 1952 to 1962, with the last B-52 produced on June 22, 1962. However, that aircraft had an FY 1961 tail number (61-040), perhaps explaining the discrepancy with Baugher. Contradictions of this type between data sources are, unfortunately, not uncommon.

It is also instructive to consider these data broken down by type of aircraft. Figure 3.3 shows the mean design age of aircraft in operation for attack aircraft, bombers, cargo aircraft, fighters, and other types of aircraft (e.g., trainers, liaison aircraft). The data are more variable when viewed by design type, but the overall pattern of growth in mean design ages is still apparent.

Between 1963 and 1988, the bomber fleet had the oldest mean design age, but the introductions of the B-1 (1985) and B-2 (1989) reduced the mean design age of that fleet. From 1992 to 1995, the small attack fleet (the A-7 introduced in 1969 and the A-10 introduced in 1971) had the oldest mean design age.

Yet another way to examine the growth in design longevity is to consider the ages at which designs were retired. The mean age of aircraft designs that were in their last year of operation in a given year is shown in Figure 3.4. For instance, the last year of operation for both the C-140 and the SR-71 was 1990. Both were introduced in 1966, making 1990 the 25th year of operation for each design. Again, the pattern of design longevity is clear. There has been a persistent, and remarkably consistent, growth in design longevity over the past 50 years.

As an illustration of the way mean design ages of aircraft in their last year of operation have evolved since World War II, the mean age of aircraft designs whose last year of operation was between 1941 and 1945 was 4.5 years. The mean age of aircraft designs whose last year of operation was between 1985 and 1994 was 19.9 years. The age of designs leaving the inventory has risen at a rate of about 3.5 years per decade.

Figure 3.3
Mean Age of Aircraft Designs in Operation, by Year and Type of Aircraft

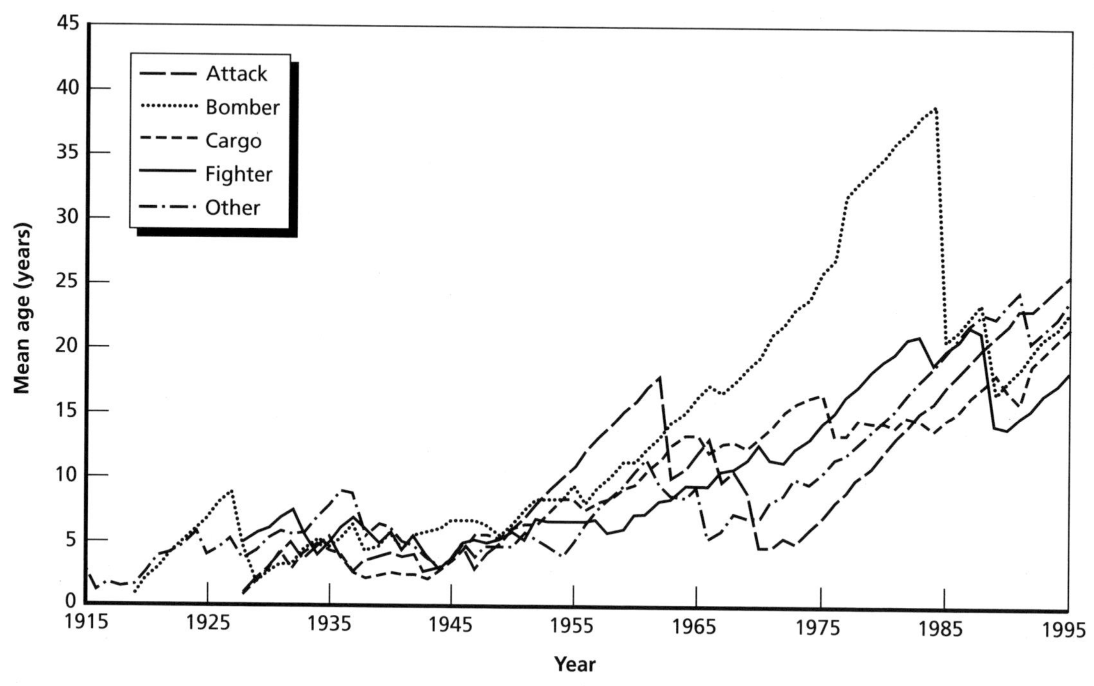

RAND *TR740-3.3*

Figure 3.4
Mean Age of Aircraft Designs in Last Year of Operation

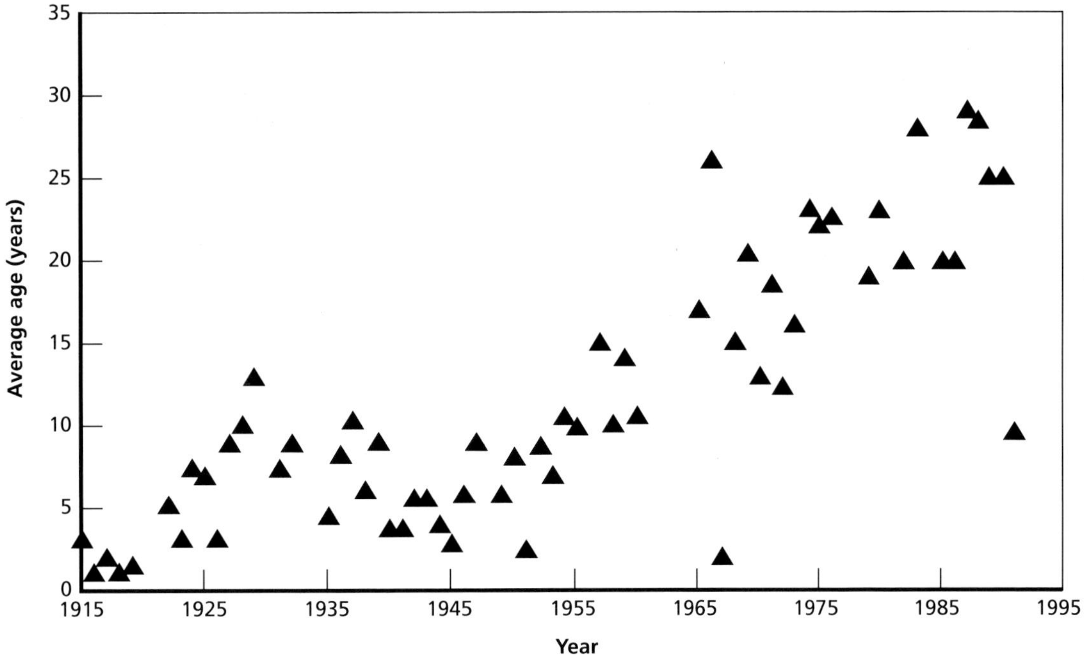

RAND *TR740-3.4*

Figure 3.5 uses the same data as Figure 3.4, but plots the actual data rather than just the annual means. The cluster of shapes in the lower left-hand part of Figure 3.5 reflects large-scale retirements of a number of designs of the same age during the early years of military aviation and again during World War II.

Figure 3.6 shows the number of aircraft design introductions and the number of those designs retired after five or fewer years of operation, aggregating the data into five-year blocks. The blocks with the most new designs introduced (1916–1920 and 1941–1945) also had the highest proportion of those designs that were in operation for five or fewer years. In the 1936–1940 block, more than half of the designs introduced (13 of 23) were also operated for five or fewer years, presumably because those designs were found to be inadequate when the demands of World War II arose.

The general correlation of many design introductions with a high proportion of designs being retired quickly could be interpreted in several ways. One interpretation is that introducing many designs allows creativity and risk-taking, only some of which result in success (although those successful designs might be highly valuable to the Air Force). Another interpretation is that there was pressure to do something quickly during those periods with many introductions, and some of the designs introduced turned out to be immature or otherwise unsuitable.

Whatever the cause, these short-lived designs have little effect on the pattern of growth in design age that we observed. Clearly, they do not affect the oldest design then in operation. The retirement of a short-lived design also has little effect on the average age of designs

Figure 3.5
Age of Aircraft Designs in Last Year of Operation

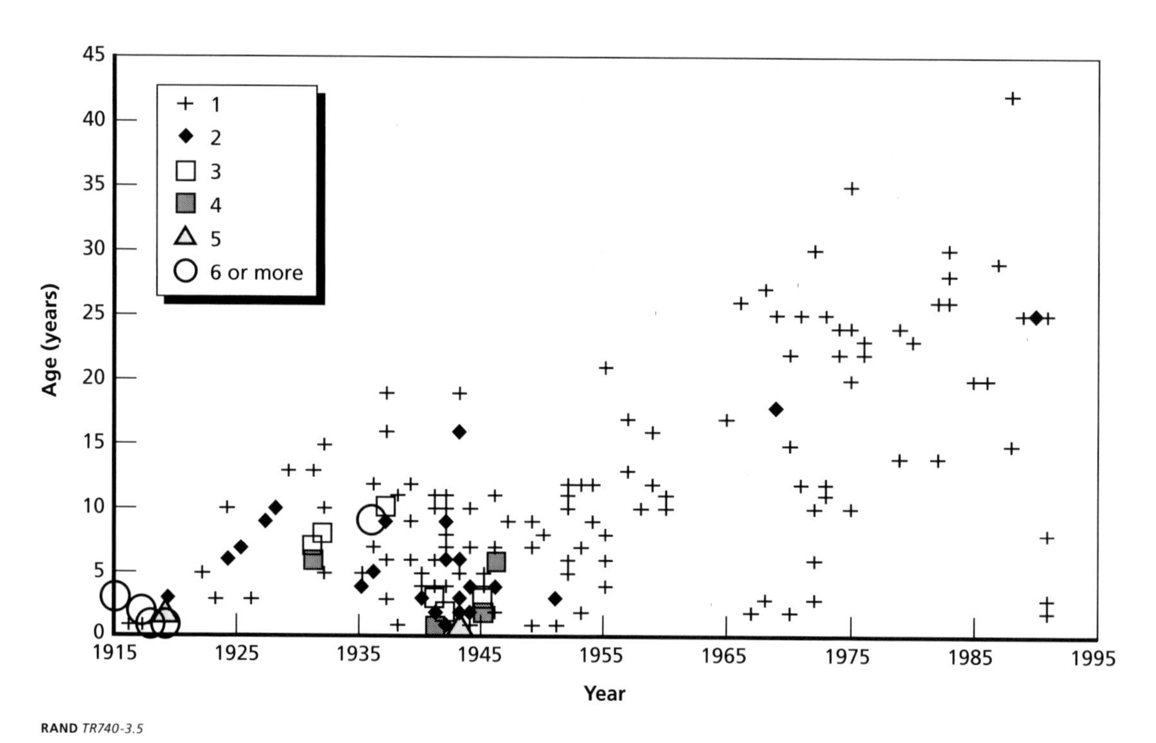

RAND *TR740-3.5*

Figure 3.6
Aircraft Design Introductions and Number of Designs Retired in Five or Fewer Years

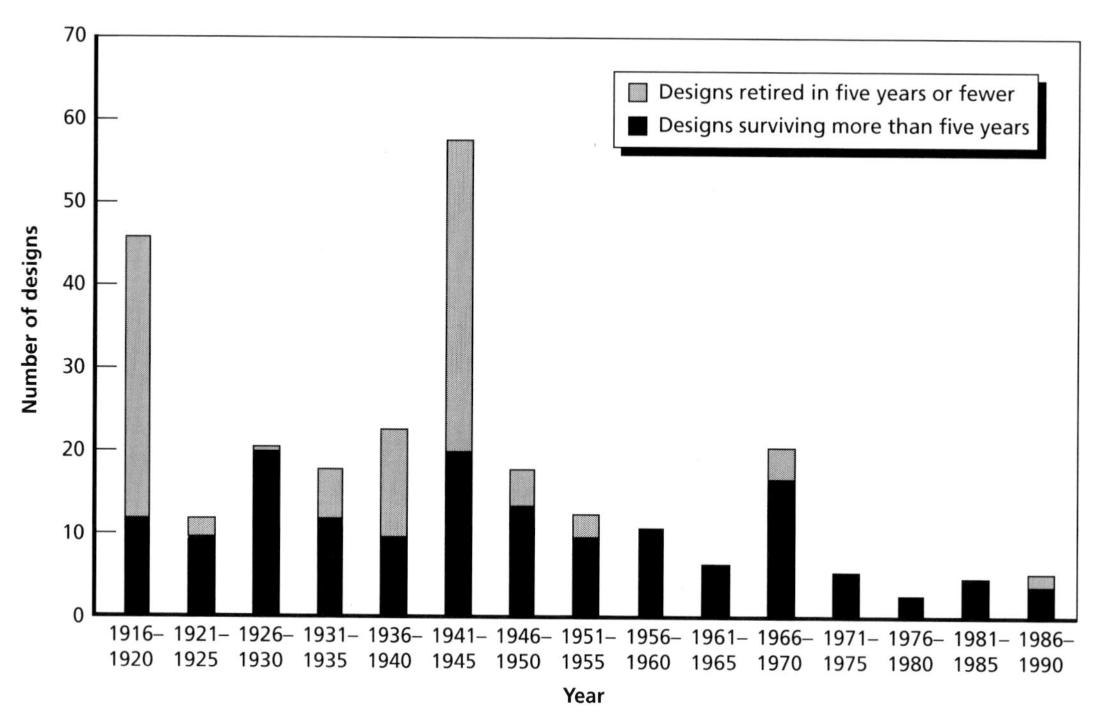

RAND *TR740-3.6*

in operation each year. In the end, the real effect of these short-lived designs may have been to serve as test aircraft to improve aviation technology, indirectly leading to better, longer-serving designs in later years.[4]

---

[4] In reviewing an earlier version of this report, our late colleague Hy Shulman pointed out that aircraft design experimentation has largely been shifted from the military services to the Defense Advanced Research Projects Agency and the National Aeronautics and Space Administration, which may help to explain the decrease in shorter-lived Air Force aircraft designs in more recent years.

CHAPTER FOUR

Discussion

The challenge to the USAF posed by aging aircraft designs did not emerge suddenly. Indeed, that challenge has confronted the Air Force since its post–World War II beginnings. The Air Force's earliest experience was with designs that were operated, on average, five years before being replaced. By the mid-1960s, the average age of designs in operation had risen to 10 years. By the mid-1990s, that average had reached 20 years.

Since the founding of the Air Force, every airman has served with designs that would reach an "unprecedented" age before finally being retired. As shown in Figure 4.1, the average retirement age of aircraft designs and the average age of designs in service have both grown roughly between three and three and one-half years per decade since the founding of the Air Force.

Keeping aircraft designs in service for a long time is not necessarily a bad thing. Presumably, designs serve as long as they are technologically and economically viable. During World War II, the U.S. Army Air Forces experienced a churning of designs. That fairly rapid turnover

Figure 4.1
Average Age of Aircraft Designs in Operation and at Retirement, by Decade

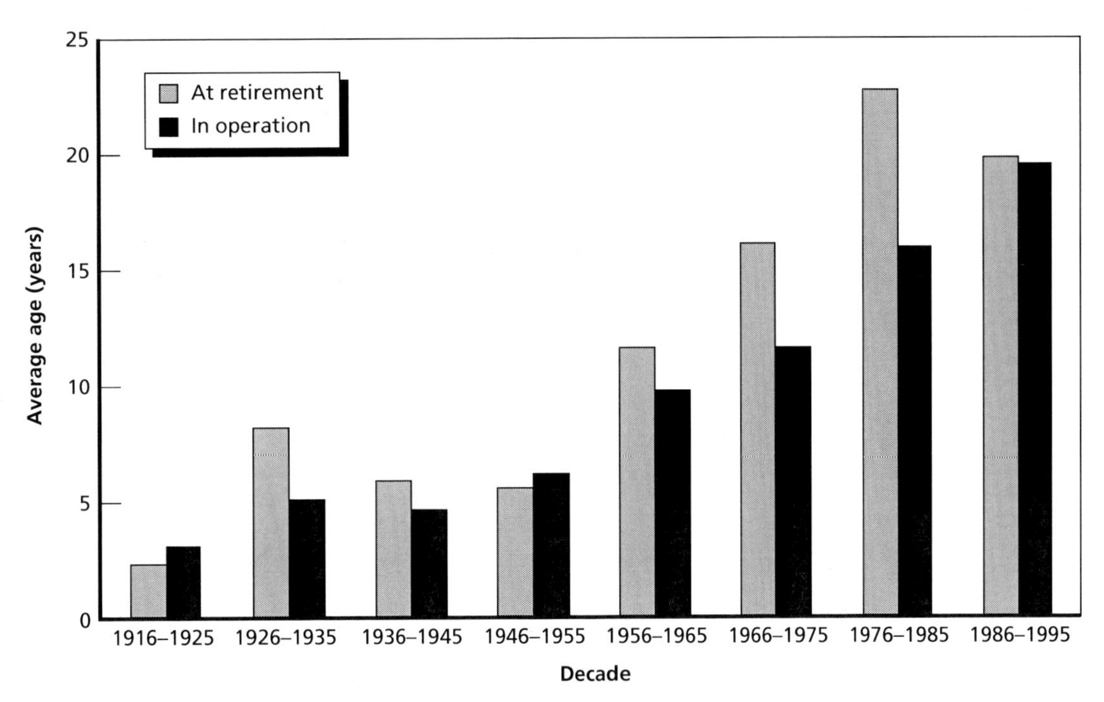

RAND *TR740-4.1*

was, at least in part, a response to the inadequacy of prewar designs to meet the war's vast challenges.[1] The World War II era was also one of rapid evolution in aviation technology, perhaps exactly because of the war's challenges. The evolution experienced in that time period tended to affect the fundamentals of aerospace design and production; most aircraft types saw new developments and turnover in designs.

Technology has continued to evolve. In recent years, however, the evolution has become more focused. Specific systems or aspects of aviation have been affected, while other aspects have remained largely unchanged. For example, stealth capability is a recent technological innovation that has been applied with effect to fighter and bomber aircraft. The basic technology of cargo aircraft, on the other hand, has not changed radically in many years.

Since the end of World War II and the formation of the Air Force as an independent military service, there has been a pronounced trend for the Air Force to keep aircraft designs in operation for ever-longer periods of time. While there is reason to believe that older aircraft are more costly to maintain, the Air Force has shown remarkable skill in its ability to manage the ever-increasing age of its aircraft designs.[2] Throughout its history, the Air Force has found ways to keep older aircraft both relevant to the mission and operating safely. The mean age of aircraft designs currently in operation is at an all-time high. But the same statement could have been made at almost any time in the history of the Air Force.

---

[1] Reviewing an earlier version of this report, Michael W. Wynne suggested that recent conflicts in Iraq and Afghanistan have caused analogous churn and development in the military's usage of UAVs. UAVs are not examined in this report.

[2] Dixon, 2006, provides a review of the literature concerning the maintenance costs implications of aging aircraft. See also Pyles, 2003.

APPENDIX A

Aircraft Designs Examined for This Report

This appendix lists the designs identified in AAS and AAW that we examined for this report. Designs are organized into five tables: attack aircraft, bomber aircraft, cargo aircraft, fighter aircraft, and other aircraft. Designs are listed in their tables by first reported year of operation.

Aircraft design designations have sometimes been reused by the Air Force. Where this has occurred, we have appended a letter designation, e.g., A-10a (1939–1941) and A-10b (1971–present). Where the same design was present in the histories under more than one designation (e.g., P-51 and F-51), we have selected one designation and used it consistently throughout.

Table A.1 presents the attack aircraft designs examined for this report. Attack aircraft are aircraft intended for direct attack on ground-based targets, usually in support of ground troop action. In all the tables in this appendix, cases in which the last report was in 1995 are in boldface, to differentiate aircraft designs that were still in operation as of the writing of AAS and AAW.

Table A.1
Attack Aircraft Designs Examined

Design	Alternative Nomenclature	First Report	Last Report	No. of Reports
A-3		1928	1936	3
A-12		1933	1942	4
A-17		1936	1941	12
A-19	YA-19	1936	1940	1
A-18		1937	1942	7
A-10a	OA-10a	1939	1941	1
A-20	DB-7, F-3, P-70	1940	1946	47
A-24	RA-24	1941	1945	14
A-29		1941	1943	5
A-35		1942	1943	3
A-36		1943	1945	4
A-25		1944	1945	1
A-26	A(B)-26, A/RA-26, FA(RB)-26, FA-26	1945	1969	29
A-1		1963	1972	10

Table A.1—Continued

Design	Alternative Nomenclature	First Report	Last Report	No. of Reports
A-37	OA-37	1967	1991	12
A-7		1969	**1995**	20
A-10b	A/OA-10, OA-10b	1971	**1995**	28

We also display, where applicable, the alternative AAS/AAW nomenclature used to describe a design. We also show the number of "reports" of the design—for example, an AAS squadron or AAW wing has the design (in one of its nomenclatures) in use. ("Reports" is not the same as a tally of squadrons and wings using the design. It is possible, for instance, that a specific wing shows both the DB-7 and the F-3, but we count both of those as A-20 variants. Nevertheless, such a case would count as two A-20 "reports": one for the DB-7 and one for the F-3.)

Table A.2 presents the bomber aircraft designs examined for this report. Bombers are aircraft intended for attack on ground targets, usually deep within enemy territory.

Table A.2
Bomber Aircraft Designs Examined

Design	Alternative Nomenclature	First Report	Last Report	No. of Reports
Caproni		1919	1928	1
O/400	HP O/400, O-400	1919	1928	2
B-3		1928	1937	7
B-5		1928	1937	4
B-6		1928	1937	6
B-4		1929	1937	4
B-2a		1931	1935	2
B-12	YB-12	1932	1942	11
B-9	YIB-9	1932	1936	1
B-10	YB-10	1935	1942	13
B-17	B/RB-17, B-17/F-9, DB-17, F-9, FB-17, RB-17, SB-17, VB-17, YB-17	1935	1955	115
B-18		1935	1944	48
B-25	F-10	1936	1946	41
B-23		1939	1941	3
B-24	C-109, F-7	1941	1946	55
B-26	B/RB-26, RB-26, WB-26	1941	1966	49
B-30	LB-30	1941	1942	9
B-34		1941	1944	4

Table A.2—Continued

Design	Alternative Nomenclature	First Report	Last Report	No. of Reports
B-29	B/RB-29, ERB-29, F-13, KB-29, LB-29, RB-29, SB-29, TB-29, WB-29	1944	1959	124
B-52		1946	**1995**	45
B-36	GRB-36, RB-36	1948	1959	25
B-45	RB-45	1949	1958	8
B-50	KB-50, RB/KB-50, RB-50, RB-66, WB-50	1949	1970	29
B-47	DB-47, E-47, EB/RB-47, EB-47, EB-66, ERB-47, RB-4, RB-47, WB-47, YRB-47	1953	1974	64
B-57	EB-57, RB-57	1954	1976	23
B-66	WB-66	1956	1970	14
B-1		1985	**1995**	6
B-2b		1989	**1995**	1

Table A.3 presents the cargo aircraft designs examined for this report.. Cargo aircraft are aircraft intended for movement of personnel and supplies (including fuel—we have subsumed the KC-135 tanker aircraft under the "C-135" rubric and the KC-10 tanker aircraft under the "C-10" rubric).

Table A.3
Cargo Aircraft Designs Examined

Design	Alternative Nomenclature	First Report	Last Report	No. of Reports
C-8		1928	1936	1
C-26		1932	1935	1
C-27a		1935	1937	2
C-33		1936	1942	8
C-39		1938	1943	12
C-40		1939	1942	3
C-34		1940	1941	1
C-45	F-2, RC-45, RF-2	1941	1957	26
C-47	AC-47, C/AC/HC-47, C/TC/VC-47, C/VC-47, DC-3, EC/HC-47, EC-47, FC-47, HC-47, SC-47, TC-47, VC-47	1941	1975	105
C-50		1941	1942	2
C-53		1941	1946	9
C-46	TC-46	1942	1968	61
C-49		1942	1943	4

Table A.3—Continued

Design	Alternative Nomenclature	First Report	Last Report	No. of Reports
C-56		1942	1942	2
C-60		1942	1944	5
DC-2		1942	1942	2
C-36		1943	1943	1
C-54	C/SC-54, HC-54, RC-54, SC-54, VC/C-54, VC-54	1943	1972	56
C-57		1943	1946	2
C-78	UC-78	1943	1946	6
C-87		1943	1944	3
C-64	UC-64	1944	1945	6
C-61		1945	1946	1
C-82		1946	1954	28
C-74		1948	1955	2
C-119	AC-119, C/AC-119	1949	1973	36
C-99	XC-99	1949	1949	1
C-122	YC-122	1950	1955	2
C-124		1951	1974	28
C-118	C/VC-118, EC-118, VC-118	1952	1975	12
C-97	HC-97, KC-97	1952	1969	29
C-121	EC-121, RC-121, TC-121, VC-121	1955	1976	23
C-123	AC-123, C/UC-123, UC-123, VC-123	1956	1975	20
C-130	AC-130, C/MC-130, DC-130, EC-130, HC-130, MC-130, RC-130	1956	**1995**	96
C-131	VC-131	1956	1979	13
C-135	EC/KC-135, EC-135, KC-135, NKC-135, RC-135, TC-135, VC-135, WC-135	1957	**1995**	105
C-133		1960	1971	3
C-141		1965	**1995**	22
C-137	EC-137, VC-137	1966	**1995**	4
C-140	VC-140	1966	1990	6
C-6	VC-6	1966	1985	3
C-7		1967	1972	5
C-9	VC/C-9, VC-9	1968	**1995**	9

Table A.3—Continued

Design	Alternative Nomenclature	First Report	Last Report	No. of Reports
C-5		1969	**1995**	9
C-12		1976	**1995**	16
V-18	UV-18	1979	**1995**	2
C-10	KC-10	1981	**1995**	10
C-20		1983	**1995**	4
C-21		1984	**1995**	9
C-22		1984	1991	1
C-25	VC-25	1990	**1995**	1
C-29		1990	1991	1
C-27b		1991	**1995**	2

Table A.4 presents the fighter aircraft designs examined for this report. Fighter aircraft are intended for attack on other aircraft. Note that we have categorized the F-15E as a separate design from the F-15A/D (which is labeled "F-15").

Table A.4
Fighter Aircraft Designs Examined

Design	Alternative Nomenclature	First Report	Last Report	No. of Reports
JN-6	JN	1917	1929	8
Sopwith:F.1		1917	1919	4
AVRO:504 K		1918	1918	2
DH-4	DB-4, DN-4	1918	1932	28
Salmson:2		1918	1919	3
Sopwith:FE.2		1918	1919	1
Spad:VII		1918	1919	3
Spad:XI		1918	1919	2
Spad:XIII		1918	1922	7
Fokker:VII	Fokker D-VII, Fokker:VII (D VII)	1919	1925	4
PW-5		1919	1924	1
SE-5		1919	1927	6
GA-1		1921	1923	2
PW-8		1924	1926	3
COA-1		1925	1931	1
P-1		1925	1932	3
P-12		1925	1943	18

Table A.4—Continued

Design	Alternative Nomenclature	First Report	Last Report	No. of Reports
P-2		1926	1931	1
P-3		1926	1931	1
P-5		1926	1931	1
PW-9		1926	1931	3
P-6		1928	1938	8
P-16		1932	1935	3
P-26		1934	1942	14
P-30	PB-2	1934	1939	4
P-35		1934	1942	17
P-36		1938	1943	29
P-37	XP-37, YP-37	1938	1940	6
P-40		1939	1947	73
P-43	YP-43	1939	1943	10
P-38	F-4a, F-5a	1941	1946	62
P-39	P-400	1941	1952	56
P-66		1941	1941	1
PT-17		1941	1941	3
PT-47		1941	1941	1
P-47	F-47, P(F)-47	1942	1952	96
Spitfire		1942	1945	9
Beaufighter		1943	1945	2
P-51	F/RF-51, F-51, F-6, P(F)-51, P(F)-51/F-6, RF-51	1943	1954	139
P-61	F-15a, F-61, P(F)-61, RF-61, YP-61	1943	1950	23
P-63		1943	1944	1
P-59	YP-59	1944	1945	2
P-80	F-80, FP(RF)-80, FP-80, P(F)-80, RF-80	1945	1957	75
P-84	F-84, P(F)-84, RBF-84, RF-84	1947	1971	74
F-82		1948	1952	15
P-86	F-86, RF-86	1949	1965	95
F-94		1950	1960	21
F-89		1951	1960	17
Meteor8		1951	1951	1
F-100	QF-100	1954	1983	76

Table A.4—Continued

Design	Alternative Nomenclature	First Report	Last Report	No. of Reports
F-102	F/TF-102, TF/QF/TQM-102	1956	1983	27
F-101	RF-101	1957	1982	27
F-104	F/TF-104	1958	1983	8
F-105		1958	1980	43
F-106		1959	1987	16
F-4	RF-4	1963	**1995**	133
F-5		1965	1989	6
F-111	EF-111, FB-111	1968	**1995**	34
F-15	F/TF-15, TF-15	1971	**1995**	45
F-16		1971	**1995**	46
F-117		1989	**1995**	3
F-22	YF-22	1989	**1995**	1
F-23	YF-23	1989	1991	1
F-15E		1992	**1995**	1

Table A.5 presents the other aircraft designs examined for this report. This category includes designs used primarily for liaison, intelligence gathering, communication support, training, and noncombat functions. Also included here are those designs that could not be readily assigned to any other category, such as the earliest military aircraft, each of which served in almost every role.

Table A.5
Other Aircraft Designs Examined

Design	Alternative Nomenclature	First Report	Last Report	No. of Reports
Burgess:F		1913	1915	1
Burgess:H		1913	1915	1
Burgess:I	Burgess:I (Scout)	1913	1915	1
Burgess:J	Burgess:J (Scout)	1913	1915	1
Curtiss:D		1913	1915	1
Curtiss:E		1913	1915	1
Curtiss:H		1913	1915	1
Martin:TT		1913	1915	1
Wright	Wright:B, Wright:C, Wright:D (Scout)	1913	1915	3
Jenny	JN, JN-2, JN-3, JN-4	1915	1924	12
D-5		1916	1917	1

Table A.5—Continued

Design	Alternative Nomenclature	First Report	Last Report	No. of Reports
H-2		1916	1917	1
H-3		1916	1917	1
JN-5	Twin JN	1916	1917	1
N-8		1916	1916	1
R-2		1916	1917	1
R-Land		1916	1917	1
Sturtevant Trainer		1916	1917	1
AR-1		1917	1918	1
R-4		1917	1919	1
Sopwith:Scout		1917	1917	1
Breguet:14		1918	1919	3
Nieuport:27		1918	1918	2
Nieuport:28		1918	1918	3
Nieuport:80		1918	1918	1
S-4		1918	1918	3
Sopwith:1		1918	1918	1
Albatros:V	Albatros:V (D V)	1919	1919	1
DFW:C V		1919	1919	1
Halberstadt:C IV		1919	1919	1
Halberstadt:C V		1919	1919	1
Hannover:C L III		1919	1919	1
LB-1	XLB-1	1919	1927	2
LVG:VI	LVG:VI (C VI)	1919	1919	1
N-9		1919	1925	1
O-2a		1919	1937	9
Orenco D		1919	1924	1
Pfalz:III	Pfalz:III (D III)	1919	1919	1
Pfalz:XII	Pfalz:XII (D XII)	1919	1919	1
Roland:VI	Roland:VI (D VI)	1919	1919	1
Rumpler:C		1919	1919	1
S-1		1919	1931	2
S-9	RS-9	1922	1937	1
LB-5	XLB-5	1923	1932	5

Table A.5—Continued

Design	Alternative Nomenclature	First Report	Last Report	No. of Reports
O-1a		1925	1936	4
O-5		1925	1931	1
O-6	XO-6	1925	1932	1
OA-1		1925	1931	2
Y-8		1925	1932	1
LB-7		1928	1932	3
O-11		1928	1936	2
O-25		1928	1939	3
O-31	O-43, O-46, YIO-31, YIO-35	1928	1943	17
O-38		1928	1943	7
O-39		1928	1936	3
O-40	YIO-40	1928	1936	3
OA-2		1928	1936	1
LB-6		1929	1937	3
O-13		1930	1936	1
O-19		1931	1939	7
OA-3		1931	1941	3
O-27		1932	1936	3
OA-4		1932	1941	5
OA-5	YOA-5	1932	1937	2
OA-9		1937	1942	4
Y-10		1937	1940	1
BC-1		1938	1938	1
BT-9		1938	1941	1
G-1	YG-1	1938	1940	1
O-47		1938	1944	12
OA-8	YIOA-8	1939	1941	2
O-51	YO-51	1940	1941	1
BC-2	BC-1A	1941	1942	1
BT-13		1941	1944	4
O-49	L-1	1941	1946	15
O-50	YO-50	1941	1941	1
O-52		1941	1943	7

Table A.5—Continued

Design	Alternative Nomenclature	First Report	Last Report	No. of Reports
O-59	L-4, L-59	1941	1949	17
L-3		1942	1944	3
L-5		1942	1953	18
L-6		1942	1944	3
G-4	CG-4	1943	1945	4
L-2		1943	1943	1
Q-8	PQ-8	1943	1943	1
T-11	AT-11	1943	1952	24
T-17	AT-17	1943	1943	2
T-23	AT-23	1943	1943	1
T-6	AT-6	1943	1949	17
G-5	TG-5	1944	1945	1
Oxford		1944	1944	1
L-13		1947	1952	6
T-33	AT-33, ET-33, T/AT-33, T/WT-33, T-6, T-7, WT-33	1947	1988	41
T-7	AT-7, RT-7	1947	1953	4
G-15	CG-15	1949	1951	2
G-18	YG-18	1949	1951	2
L-16		1952	1953	1
L-20		1952	1955	7
SA-16	HU-16, SA/HU-16	1952	1969	23
T-28	RT-28, T/AT-28, T-43, YAT-28	1953	**1995**	12
L-28	L-28/U-10, U-10	1962	1973	10
O-1b		1963	1973	4
T-37	YAT-37	1964	**1995**	24
R-71	SR-71	1966	1990	4
T-29	VT/T-29, VT-29	1966	1975	6
T-39	CT-39, T(CT)-39, NT-39	1966	**1995**	15
U-2	TR-1, U-2R/TR-1, WU-2	1966	**1995**	15
U-3		1966	1968	2
U-4		1966	1979	3
U-6		1966	1967	2
O-2b		1967	1986	7

Table A.5—Continued

Design	Alternative Nomenclature	First Report	Last Report	No. of Reports
T-38	AT-38, CT-38	1969	**1995**	43
U-21	YQU-21	1969	1970	1
V-10	OV-10	1969	1982	4
U-22	QU-22	1970	1972	2
T-41		1972	**1995**	8
E-4		1974	**1995**	3
T-22		1974	1988	1
E-3		1977	**1995**	7
T-1		1992	**1995**	2
T-43	CT-43	1992	**1995**	3

APPENDIX B

Results After Removing One-Report Designs

Some of the designs presented in this analysis were in the Air Force only briefly, perhaps as test aircraft. The F-23, for instance, appears in the data between 1989 and 1991, but it was ultimately rejected in favor of the F-22. We wanted to evaluate whether aircraft with such limited duration or use influenced any of our central findings.

We did not have an explicit flag in the data that a design was only used for test purposes. As a proxy, however, we removed any aircraft for which we had only one report in AAS and AAW. We removed, in other words, any aircraft with the value of "1" in the rightmost column of Tables A.1–A.5. Some of the aircraft removed by this algorithm are clearly test aircraft, such as the F-23. Others, however, are "real" aircraft, such as the B-2 (which, heretofore, has been stationed only at Whiteman Air Force Base—so there is only one report for that design in our data).

Figure B.1 is a modified version of Figure 2.1, showing aircraft designs in operation by year, removing one-report designs.

Figure B.1
Aircraft Designs in Operation, by Year (Only Designs with More Than One Report)

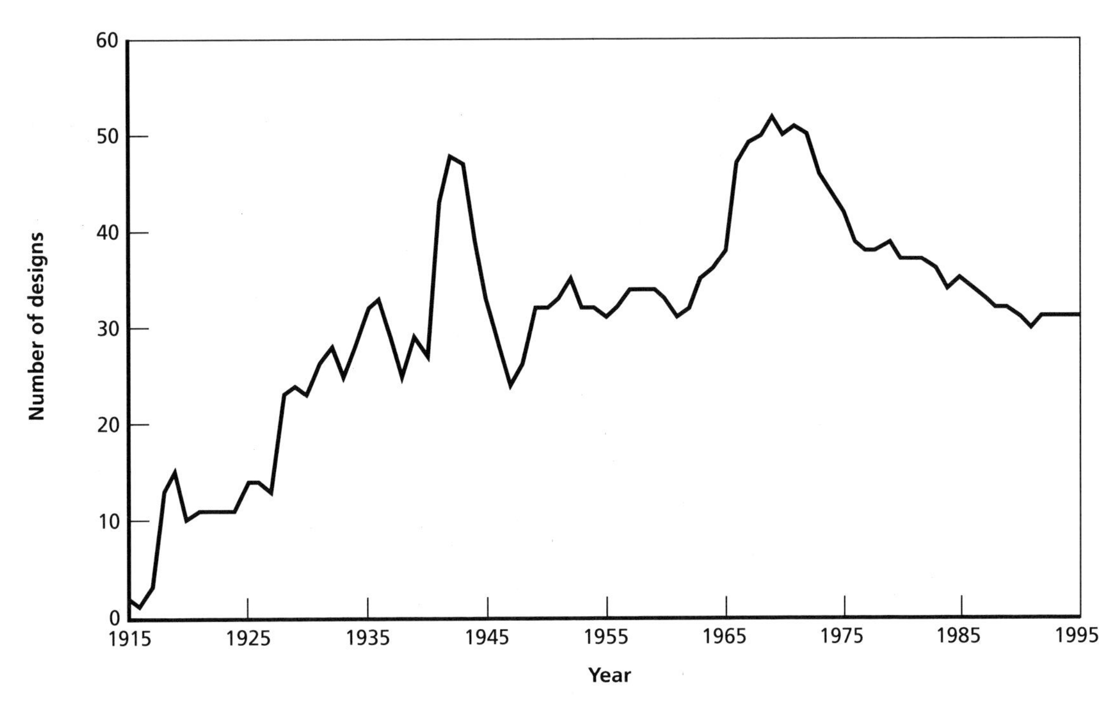

RAND *TR740-B.1*

Figure B.1 resembles Figure 2.1 in that it has two peaks, one proximate to World War II, the other proximate to the Vietnam War. Fewer designs are observed early in the data, however; there were proportionally more one-report designs in the earlier years of military aviation.

Figure B.2 is the analog to Figure 3.1, showing maximum, median, and mean age of operated designs by year, but only for designs with more than one report.

Figure B.2's maximum line is the same as Figure 3.1's World War II line; the Air Force's oldest designs were all multiple-report aircraft. Likewise, the median and mean lines show the same basic pattern of increases as those in Figure 3.1.

Our conclusion is that the results are not meaningfully different, whether or not one chooses to include one-report designs in the analysis.

Figure B.2
Maximum, Median, and Mean Age of Designs in Operation, by Year (Only Designs with More Than One Report)

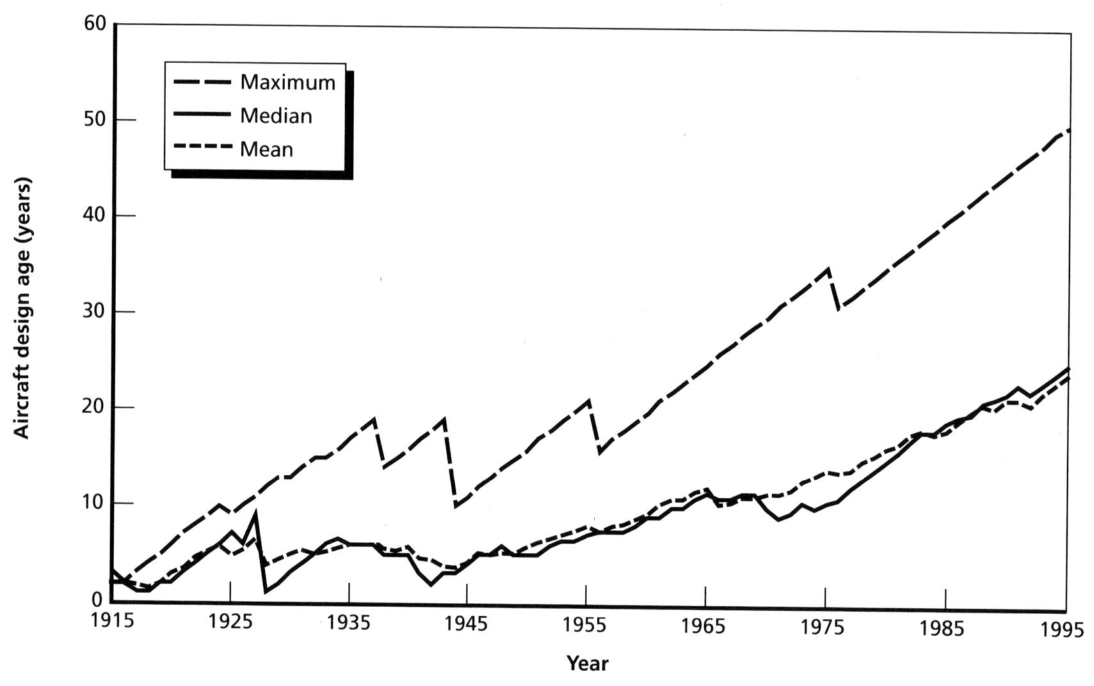

RAND *TR740-B.2*

References

Baugher, Joseph F., "USAAS-USAAC-USAAF-USAF Aircraft Serial Numbers—1908 to Present," last updated December 6, 2008. As of December 13, 2008:
http://home.att.net/~jbaugher/usafserials.html

Boeing Integrated Defense Systems, "B-52 Stratofortress: 50 Years of Exceptional Service Span Cold War to Enduring Freedom," 2002. As of June 21, 2009:
http://www.boeing.com/defense-space/military/b52-strat/b52_50th/index.html

Dixon, Matthew C., *The Maintenance Costs of Aging Aircraft: Insights from Commercial Aviation,* Santa Monica, Calif.: RAND Corporation, MG-486-AF, 2006. As of August 6, 2009:
http://www.rand.org/pubs/monographs/MG486/

Futrell, Robert Frank, *Ideas, Concepts, Doctrine: Basic Thinking in the United States Air Force, 1961–1984,* Vol. 2, Maxwell Air Force Base, Ala.: Air University Press, 1989

Pyles, Raymond A., *Aging Aircraft: USAF Workload and Material Consumption Life Cycle Patterns,* Santa Monica, Calif.: RAND Corporation, MR-1641-AF, 2003. As of August 6, 2009:
http://www.rand.org/pubs/monograph_reports/MR1641/

U.S. Air Force Historical Research Agency, *Active Air Force Wings as of 1 October 1995,* Washington, D.C.: Air Force History and Museum Program, 1998.

———, *USAF Active Flying, Space, and Missile Squadrons as of 1 October 1995,* Washington, D.C.: Air Force History and Museum Program, 1998.

United States Air Force Historical Studies Office, "The Lineage of the United States Air Force," 2009. As of June 21, 2009:
http://www.airforcehistory.hq.af.mil/PopTopics/lineage.htm

Worden, Mike, *Rise of the Fighter Generals: The Problem of Air Force Leadership 1945–1982*, Maxwell Air Force Base, Ala.: Air University Press, March 1998. As of December 2, 2008:
http://permanent.access.gpo.gov/websites/dodandmilitaryejournals/www.maxwell.af.mil/au/aul/aupress/books/worden/worden.pdf